Introduction

Red deer are still living wild in a few parts of Britain – in the Lake District and in the New Forest; on Exmoor and in Scotland. They also live in a semi-natural state in some parks from which they cannot escape.

And so, in autumn the stags (male deer) can easily be seen fighting among themselves for possession of the hinds (female deer). Only the stags fight, as they have the antlers to do so; for a time they become quite unlike the docile animals pictured *above*.

They challenge each other by calling or 'belling'. If two stags do fight the matter out, they take up position a few feet apart, then approach and lock antlers. They struggle, twisting and pushing until the loser eventually grunts, turns tail and runs away, leaving the herd of hinds to the victor – called by humans the 'master stag'.

Not that the hinds seem to care much. They would not have stopped grazing to watch the battle!

And indeed, as a battle it hardly adds up to a real fight. Little blood if any is lost. The sharp points of the antlers are not used for gouging. The strength of the animals' necks, twisting the opponent, decides the contest.

So what exactly are the antlers for? Why are they sharp? For defence against other, predatory animals? If so, why does the stag shed them in March and leave itself defenceless for some months?

Many pictures have been painted of stags heroically fighting off wolves or hounds with their antlers. Is that the right picture, or have we got the way that they behave wrong? And what about other animals? Are we looking at them in the right way? In this **WATCH Book** we find out.

Starting with wasps

Almost everybody has been stung by a wasp at some time or other. The sting is a good defence against interference – we leave them alone.

That sting is also a vital tool in the wasp's life. The adult wasps we see are vegetarians, feeding on nectar and sweet liquids – we notice them mostly in the autumn when they come to 'drink' at rotting fruit and jam. But they are active in the garden all summer, catching other insects, stinging them, and carrying them back dead or stunned, to feed their grubs in the nest. Wasps are relatives of bees, and build elaborate nests.

Unlike bees, however, the worker wasps (that you see flying) die in the autumn; only the queen wasps survive. Perhaps it is their imminent death, or perhaps it is the fact that their job is done, that 'triggers' off a change in the wasps' behaviour – so that they plague the jam pots.

And we can start by asking some very simple questions: *how* does the wasp home in to the jam?

It will be no bad thing to try to find the *simplest* answers to such questions, as we will discover in this book.

The **wasp nest** opposite *has been dug from the ground but you will often see them in more accessible places (*below*); notice the patterning of the surface. At* bottom, *a solitary* **sand wasp**, *this one about 2·5 cm long.*

Look for a wasp nest. That shown *opposite* was built underground; all that would be visible would be an entrance hole. But they can also be found between roof rafters (*right*).

Look for wasps actively hunting in the garden. Often they can be watched while they seize hoverflies in mid air. They sting them, and fall to ground. There they strip the body of wings and head and carry the rest to the nest. How do they locate their prey?

If you can find a nest, put some jam on a saucer outside it. Do the wasps locate the jam – if so, how? Do they fly straight to it, or do they circle around it, moving closer at each turn until they finally land?

Does one wasp, the pioneer, go back to the nest and return with others; if so, is that evidence that they can 'tell' other wasps of a food supply?

Wasps belong to a large group of insects which includes ants and bees. Many of them are 'social' – they live in large colonies and build elaborate nests. Others, however, are solitary. The sand wasp (*left*) is a solitary wasp; it stings its prey and drags it back to its burrow, dug into sandy soil. When the prey has been pulled underground, the wasp lays eggs near it; when they hatch, the grubs feed on the prey (here it is a caterpillar).

Look for other small 'mason' bees and wasps with a similar way of life. An old wall is a good place to watch for them – they burrow into the loose mortar. They can also be seen on dry sunny banks.

It is interesting to observe so much *variety* in the families of the animal world. Social or solitary; and the food eaten – they all give each wasp a different character. But that does not tell us how the wasp locates its prey!

Seeing

Being able to see is very important for us humans. We take it for granted, and it is easy to assume that the wasp 'saw' the hoverfly and the jam.

But it could also have smelt the jam from far away, and flown towards it locating it by smell, until it could 'see' it at close range. That kind of answer would explain the ever-narrowing circles it flew in.

Wasps have eyes, but as you can see from the photograph of a near relative *above*, the eye is not at all like ours. There is no single pupil; and if it were to be enlarged further, you would see that it is a *compound* eye – made of a large number of separate lenses. Do those multiple lenses give a wasp the same picture of the world as our eye?

Before we can enquire about that, we need to decide what picture our eyes give us. They tell us about light (and dark); colour; movement – is there anything else? If a wasp cannot see with the clarity that our eyes give us, which of those features would be more important than the others? How does it catch its food – would detection of movement be valuable?

We need to ask questions like that to discover how the animal world behaves. Do animals see the same colours as us humans? Or even see the same kind of light?

The **hornet** *(a relative of the wasp) shows its compound eyes in close up* opposite. Below, *the* **ultraviolet pattern of a white flower** *seen by a bee, and on the* right *a sketch giving an idea of the kind of* **sky map** *made by* **polarised light**.

Our own eyes can detect the colours of the rainbow. We do not yet know what colours wasps can see, but we know that bees *cannot* see red. To them, a red flower is greyish.

However, they can see a 'colour' we cannot; below the violet is another one – and that has a fascinating result. To a bee a 'white' flower is rarely white. Many white flowers have 'ultraviolet' stripes, which guide the bee to the nectary: one such pattern is shown *above left*.

That is not all. We believe that bees and maybe birds, too, can see a *kind* of light we are insensitive to. It is known as 'polarised light'.

Light reaches us from the sun in 'waves' – if 'seen' head on, those waves are in every direction. However, dust in the air filters out some of these directions, so that the light arrives 'polarised'. What is filtered out depends on where the sun is in the sky – but the result is that clear blue sky has a 'pattern' for eyes that can see it, and the pattern can be picked up from just a small patch of clear sky. Bees and maybe birds navigate by that pattern.

Light is often not only a thing to see by. The lacewing *above* has lovely bright golden compound eyes. It comes to windows at night. Why? If its need is for light to see by, why doesn't it fly by day? Light *attracts* some animals – that is not the same as the sense of sight.

Other senses

We have seen that sight can be very different for the animal world. How about our other senses – hearing, smell, touch?

Sound is like sight in one way – many animals can detect a range that we cannot. Bats are a well-known example; they emit high-pitched squeaks (we cannot hear them), which echo back from small objects and so give the bat a 'sonar' detection system. However, they seem deaf to low notes. Other animals use sound as a major part of their lives: grasshoppers such as that *above*, for example. Grasshoppers chirp – or, at least, the males do and that 'song' helps to attract the females to them. As we will see, song also plays an important part in bird lives.

We have so few words to describe smells that it is obviously not an important sense to us, but a male moth can track down a female of the same species from two miles away – by the scent she emits. (The feathery antennae of moths seem to be used to detect smells.)

How about touch? We often forget we do have a well-developed sense of the way objects feel – we can tell whether a surface is wood or cloth by tapping it with a finger for a fraction of a second. Crabs and many other animals are covered with very sensitive 'touch' hairs.

Are we sensitive to wet? Some animals are!

The **meadow grasshopper** opposite *uses sound as a help to attract the female, while the* **male midge** below *(7× life size) is attracted by the hum of the female midge.* **Woodlice** *(*bottom*) show simple reflex reactions that you can observe.*

Look for swarms of midges, drifting in the breeze and flying back, so that the swarm is rather like a balloon tied to some 'marker' on the ground. (And the marker may be something not at all obvious to our senses!)

Can you follow the movements of one midge within the swarm?

Midges fly with a high-pitched hum – which has its use. The female also emits the sound; the males pick it up, possibly with their feathery antennae (*above*) – and home in on her.

However, does that explain how a blood-sucking biting midge or mosquito can home in to you? It could *smell* you maybe – or it could be sensitive to the *heat* of your body. Watch some mosquitos or midges. Do they land on your clothes *often*, *sometimes* or *never*? Or only on your bare skin? What would that observation suggest?

Look at a column of ants – they have rudimentary eyes which can probably only tell them 'light' and 'dark'. They use smell to recognise each other. How do they greet each other? Is it with Antmusic?

Summing it up with woodlice

Woodlice (*below*) are easy to find – all you have to do is disturb some leaf litter near a tree or lift up a stone and you will usually find some. When exposed to light, they scuttle and hide themselves away. They are attracted by two things – *dark* and *damp*. They are repelled by *light* and *dry*.

They have no choice. They steer automatically towards the first and away from the second.

That kind of involuntary reaction can be seen throughout the animal world. It explains *how* the wasp homed in to the jam – zig-zagging until it landed exactly on it. It explains how moths and lacewings kill themselves on hot light bulbs.

So, animals detect a *stimulus* (light, sound, heat, smell) with a *sense organ* (eye, ear, antenna, nose) and *react* to it – the reaction often being a *movement* to the stimulus or away from it.

But surely life is not that simple?

Drives

We humans have some simple reflexes – if we accidentally pick up a very hot plate, we cannot hold it – we drop it. We blink if something moves towards our eyes. Those reactions protect our bodies, and they are rather similar to the 'simple' reactions of the woodlice when you scatter the leaves hiding them.

However, there is more to being alive than just 'reaction'. We and all animals have *drives* – and one of the easiest drives for us to understand is *hunger*. What makes animals feed? What stops them?

The kestrel above makes a good starting point. It is feeding its nestlings, as you can see. However, out of the breeding season it catches food only for itself. And those who have kept a kestrel in captivity have discovered something puzzling: if they feed it, it will continue to 'eat' until it cannot take any more – until it is even too heavy to fly!

Something tells it that it is 'hungry'. But nothing seems to tell it that it is not!

So what happens to it in the wild? Why doesn't that kestrel catch *all* the mice in one place – so that it becomes too heavy to fly?

Another question is: does it ever stop feeding if it never 'feels full'?

In other words, does the rest of the animal world have the meal times we take for granted?

The **kestrel** opposite *and the* **mole** below *make good starting points; they are both predators but with rather different food drives. The kestrel will gorge until replete during daylight hours; but the mole has its own four-hour 'day'.*

Look for kestrels. They are common along motorways, where the long grass forms an ideal home for the voles and mice that they hunt.

How often do they dive? Hovering is an energetic way of keeping aloft; it requires non-stop rapid wing beats. A lot of energy is needed. A lot of food is needed to supply it.

Kestrels roost at night: but do you see them perched at rest during the day? If so, do they 'rest' for long?

Look at birds feeding in the garden. (In late spring and summer they are also feeding their nestlings, so their behaviour may be different).

In the winter months, they do seem to feed more in the morning than in the afternoon. You can test that by putting out food during the coldest days. Does food put out in the afternoon attract as many birds as food put out in the early morning?

Listen out for owl hoots. Do you ever hear them in the middle of the night, or only at dusk and dawn?

Many animals seem to have periods of activity, followed by periods of rest.

Look out for mole hills. If you see the earth heaving, take a note of the time and return next day. Is the mole active at the same time?

Moles are well adapted for their way of life: they dig a tunnel through good soil, and run regularly along it, catching worms and small animals that have fallen into it.

And it seems that the mole has a four-hour 'day'. It is active for four hours, rests for four, is active for four . . . night and day.

It has a regular clock of some kind, but a clock that is different from the day-night, sun-based clock that seems to control the kestrel – and also rules our own lives.

Flying bodies

The kestrel is a beautiful example of a hunting bird – in the same way that the mole is a beautiful example of a tunnelling mammal. Both of them are well adapted to suit the kind of life they lead.

And so is all animal life, when you think about it. And you recognise that fact by 'expecting' to see certain animals in certain places.

The owl is another example. Weight for weight, the barn owl *above* is probably about the same as the kestrel (they are certainly the same length, beak to tail), but they look different because of the way they live.

Owls can fly almost silently through a dark wood. Their eyes are large, to take in as much light as is possible at night. However, they can also maybe see an extra 'infrared' colour, and possibly to an owl a warm-bodied creature like a mouse actually glows.

They also have sensitive ears – the barn owl *above* could exactly locate a mouse rustling the straw in a pitch-dark barn.

The wing is a beautifully designed instrument, but the shape it is and the way it functions is different from species to species. But each wing is exactly suited to the bird's way of life. Below right, *some wing profiles to look for.*

Look at birds flying to see examples of *shape* engineered to suit the way of life of the bird.

The 'secret' of bird flight was not known two centuries ago. But then discoveries quickly followed each other, and so did other inventions such as photography, which made an explanation possible.

Bird flight relies on the use of an 'aerofoil' wing.

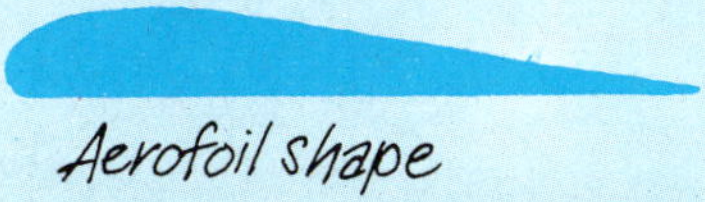

When a wing with that shape in cross section is moved through air, it creates a 'lift', enough to carry the weight of a body. However, the bird wing is a complicated design because as well as this, it has also to be the propeller – an aerofoil tilted forward to pull the bird *through* the air.

With wings outstretched and unmoving, birds glide. Narrow wings give a *fast* glide, losing height quickly. The broad wings of the owl can give a *slow* glide, losing height slowly.

Seabirds such as those on the *right* usually have 'fast' wings – which enable them to skim the waves at high speed, to catch fish. They need to be able to move fast – the fish would dive down at a sighting. But they need not be able to take off quickly.

A pheasant, a bird of similar weight, has short stubby wings – it can take off quickly – it needs to, among shrubs and trees.

How do the wings of garden birds compare with each other?

Escaping

Wings enable birds to be successful predators, hunters of other animals. So let us return to the question we have not yet answered. Why then, do they not overeat? Why do any of their prey survive?

One possible reason is easy enough to guess. If the predators are well-designed catching machines, their prey may also be well designed – for escaping. The number of ways that safety is sought is fascinating. Sometimes the behaviour of animals under attack seems strange – to us.

The grass snake *above* is feigning death. If it is alarmed, it hisses, and if handled it often lets loose a foul smelling liquid. If that does not discourage its attacker, the snake will then often adopt the posture seen above – it plays dead. How can that possibly be a defence? Well, it could be, if the predator was 'programmed' to attack only moving, and hence *live* animals. That possibility tells you much about the intricacies of animal behaviour.

There are many kinds of protection. Some animals fight back. Some hide themselves, others disguise themselves with camouflage. Many play a double cross – they mimic other, more dangerous animals (many harmless hoverflies are striped yellow and black like a wasp).

Others make it obvious that they are there – but they carry bright gaudy colours as a signal that they taste foul.

The **grass snake** *opposite is feigning death, as a final defence against a threat.* Below, *a* **camouflaged caterpillar** *and another approach to protection: the* **grotesque caterpillar** *of the 'lobster moth'. Can you work out where its head is?*

Look for examples of camouflage in the garden or down the lane. The caterpillar above is that of the orange-tip butterfly; one of the first butterflies to fly in spring.

The camouflage is good – *while the caterpillar remains still.* It would be more visible if it began to move.

And if you look closer, you will see that it is 'countershaded' – it is lighter underneath than above. So the effect of its shadow is lessened, and it does not look plump and round. That too is good camouflage.

Many animals wear patterns of colours that break up their outline. Examples range from the stripes of a tiger (which make it nearly invisible in tall grass) to the wings of many moths.

(A tiger? But surely tigers are *predators.* They too sometimes adopt camouflage the better to creep up to their prey!)

Look at the lobster moth caterpillar *below* – presumably its grotesque appearance deters birds.

Birds are commonly predators, and so many of the camouflages seen on insects, including butterflies and caterpillars, will be directed at outwitting birds. That means, of course, that the eye of the bird and what it sees of the world must be similar to our own – because we also have difficulty seeing animals that are camouflaged to confuse birds.

And it is interesting to wonder what did come first – a wasp has a sting, so birds leave it alone, recognising it by its stripes. But do all young birds try wasps, and 'learn by experience'? Can birds learn in that way?

Or is their knowledge of wasps something they inherit from their parents – as much part of their character as the shape and the colour of their feathers?

Mating

Sex is a drive as strong as hunger, and ways of mating are often complicated. Even the spawning of frogs and toads is more complex than it seems.

As a start, all toads and probably most frogs have special breeding ponds – but how those ponds are chosen and why, we do not know. Toads will make long journeys to reach them – crawling in a straight line, even climbing over walls in their path. Many frogs hibernate in the mud of ponds; yet often they travel to another pond to mate.

The animals arrive at the pond all within a few days of each other in spring so that some kind of clock must be active. The males croak, a signal to the females, and when a female arrives she is grasped by a (smaller) male, as in the picture of toads *above*. At this time the males will grasp anything – fish, other males, even a stick if you move it near them. Because the females leave after breeding and because fewer female toads arrive at the start, there is competition among the males – they push each other off the female. But one is successful; she lays her eggs and he pours his sperm over them. The eggs swell to become the familiar spawn.

Opposite, *a small male grasps the larger female* **toad** *at the start of mating.* Below, **frogs** *spawning in a pool; the males try to push each other off the females. At* bottom, *a* **male smooth newt** *(life size) with crest and courtship colours.*

Look out for those special breeding pools in spring – easy to recognise by the activity in them. The exact time of spawning seems to depend on the warmth of the spring, along with other factors.

Does the pond contain both toad and frog spawn? Frog spawn is laid in a mass, toad spawn in a string, which often becomes wound round water plants. If the two are using one pond, you will find the frogs prefer sunny shallow water, while toads use the deeper parts.

Although we always associate them with water, frogs and toads both spend most of the year on land. Frogs are found in the shelter of thick vegetation where it is damp and cool with plenty of flies. Toads, however, seek out a hole as a home base – the hole is often in a greenhouse or garden shed. They have often been known to become quite 'tame'. (We'll try to find out what tame means, later in this **WATCH Book**.)

Look also for newts, breeding in spring. The smooth newt has a crest along its back, and becomes strikingly coloured, as you can see in the picture *below*.

More mating behaviour

Intricate colour patterns and behaviour are seen throughout the animal world during mating. Even fish – and sticklebacks were among the first to be investigated – give us some surprises.

In spring, the male stickleback develops the red colour and blue-green eyes in the picture *above*. The female remains as before. The male builds a loose nest of water weed; it takes him about three days. The males then perform a zig-zag 'dance' in the water in front of a female, and when a female is attracted to a male as a result, he leads her off to his nest. She lays her eggs in it, and he empties his sperm over them. The female then swims off, but the male remains. He stands guard over the nest, fanning a stream of fresh water over the eggs. After five days or so, they hatch, but the male guards the young fish for another fortnight, even taking them into his mouth for safety when a predator approaches.

The red-coloured **male stickleback** opposite *is 'dancing' for the female.* Below, *a pair of* **butterflies** *mating; and male (with antlers) and female* **stag beetles**. *The male can be about 6 cm (2½ in.) long. The female is about the same size.*

You will of course see many animals mating in the garden. Is what you see the whole event – or just part of an elaborate ritual?

The mating of some insects seems to be a fairly rapid affair, though you will often see butterflies in mating posture, flying together – or one being carried by the other.

Dragonflies too, mate in tandem, head to tail, and they are a familiar sight over water.

Look out for snails mating; they rear up against each other.

Look out for clues to the mating of spiders. Spiders are active predators; some spin webs, but other (wolf) spiders hunt their prey over the ground, and catch it by jumping at it. You may sometimes see those hunting spiders in pairs; but rarely if ever the web-spinning spider with another. Can you guess why this may be?

Look out for stag beetles, seen *below*. The male is our largest beetle. Stag beetles are found mainly in the south of England.

Only the male has the two 'antlers' which are modified jaws.

The female can give a bite with her powerful mouth parts, but the male finds his 'antlers' difficult to move with any power. However, they can be moved a little, and they *may* be of some use in mating 'battles' with other stag beetles.

It is a remarkable parallel with the antlers of deer – because, of course, the two animals are very different in every way.

The eggs of the beetles are laid in decaying wood (as are those of many beetles). The grubs hatch, and are also interesting – they 'chatter' – making a noise which maybe warns other grubs away from their own tunnel in the darkness of the decaying wood.

The mating of birds

The courtship rituals of birds are often complicated, fascinating and easier to observe than those of other animals. The birds in the picture *above* are black grouse. At the start of the breeding season, the males meet together in the early dawn at a traditional display ground (a 'lek') and carry out a dance, with much gobbling and shaking of feathers, as you can see. The females are attracted to males in this way.

Rituals can be followed much more conveniently in the back garden! Blackbirds make a good example. The male blackbird is deep black in colour, with a bright yellow bill. The female is mottled brown, with brown bill.

Like many birds, blackbirds have a range of noises of all kinds – mating calls, but also alarm calls and distress sounds. The mating noises begin in autumn when the birds often roost near by each other. The males begin to 'chink' at dawn and often at dusk – it is a noisy '*mik mik mik*' which sometimes breaks into a noisy alarm rattle. Through that noise the dominant males, who will attract mates, seem to be recognised.

However, that is just the prelude to a fascinating aspect of birds – their *song*. When the pair is established, they begin to sing from a 'song post' near the nest site. The song is a warning to other blackbirds to keep away – and so it marks out the *territory* – the family food patch.

The **black grouse** opposite *are 'dancing' to attract the females to pair with them.* Below, *some garden* **bird songs** *reduced and simplified, in music notation; however, our ears cannot pick up many fine details of the song.*

Nightingale

Blackbird

Thrush

Some poets have also attempted to put bird song into words.

"That's the wise thrush,
He sings each song twice over"

wrote the poet Browning.
But poet Tennyson was more ambitious. He wrote the thrush's song as:

"Summer is coming,
Summer is coming.
I know it, I know it, I know it."

Listen out for the dawn chorus. What is remarkable about it is perhaps not that so many birds can be heard singing their territorial songs, but that they are singing so close together.

Yet when you analyse what you can hear, you will realise that they are all birds of different species. In other words, birds of different kinds do not seem to be in competition. *Birds only 'compete' with others of the same species.*

That law is an important one in nature; and it applies not only to birds (though they make an easily seen example).

Try to record some of the song of the blackbird – it is easy enough to do – with the family cassette recorder or music centre – just hang the microphone out of the window. You will find, when you play the tape, that the blackbird sings a set of separate *phrases*. It is known that those phrases change, so that different ones can be heard after a week, but the advantage of such a change is difficult to guess. Perhaps it is so that other blackbirds cannot 'get used' to the territorial song?

However, other birds have a much simpler song. Listen out for the repeated song of the thrush.

Listen out for robins. They are strongly territorial – and they sing all year round. If you approach a singing robin, it may fly away from you – if you follow it, it will retreat until it suddenly doubles back past or above you. It has reached the edge of its own 'territory'.

Look out for the courtship ritual of ducks on a pond; it is often flamboyant, with much splashing as rival males 'fight'.

The music notation *above* shows songs of garden birds – but much simplified, because we cannot hear parts of their song – our ears cannot register it.

More about 'clocks'

For most animals, the mating and breeding season is tightly timed. It coincides with the time of year when there is a good supply of food available to feed the youngsters. And the exceptions only seem to underline that general rule. Town pigeons can breed all year round – they lay eggs even in the winter months. However, they are 'town pigeons' because they have adapted to live in the towns – where there is plenty of food from human houses and markets – and all year round too.

We take that year for granted. We take the day for granted too – and much animal life follows that twenty-four-hour clock, being active during daylight, or at night, but rarely both (or so it seems from our viewpoint). However, we do know about exceptions (the mole is one) and indeed many animals keep to their own 'clock'.

So, let's ask a difficult question to answer. Do animals get tired – as we ourselves do? The female mallard above looks 'tired' – so it seems to us. She is 'resting'. Or is she?

Any animal that is preyed upon by another would of course be at risk if it rested when its predator was active – unless resting was part of its protection – as in the case of the camouflaged caterpillar on page 13.

Below, *two insects which reverse the pattern of behaviour followed by most of their kind. The* **December moth** *(*top right*) flies in winter. The* **bee hawk moth** *(*bottom*) flies by day – and as protection is disguised to look like a bumble bee.*

It is always interesting to look for exceptions to the general rules, and to try to find animals that do seem to be resting; and if you find them, to try to work out *why* they are.

And look out also for animals which reverse the seasonal or daily pattern of their relatives.

Look for the December moth, *right*; it flies in winter, and is the moth you see coming to windows in the cold months when other moths are hibernating, or spending the time as an inactive pupa.

Look for a bee hawk moth, *below*. Most moths are night fliers and have a well-developed sense of smell, which enables them to locate flowers at night, but this species is active by day. To protect it against birds, the bee hawk moth is disguised to look rather like a bumble-bee – it looks very similar to that bee when it is flying.

Living places

All animals rely on a food supply. Some eat plants, or parts of them. Some eat other animals, alive or dead. Plants come first, though.

Animals also need shelter – from the weather; and protection from their enemies. Some live long lives, over many years. Some animals have long stages spent immature, with a brief adult existence, maybe as short as an hour or two. But only as adults can they breed.

The lives of those animals is obviously intricately woven, as is their relationship with the plants. And just how intricate it can be is shown by the Robin's Pin Cushion, *above*. It is found on wild roses, and it is a 'gall'. A small gall wasp lays her eggs in the unopened flower buds of the rose, and when the grubs hatch, their feeding irritates the plant cells so that they produce a 'ball' of threads, at first green but which turn red as in the picture. The grubs pupate inside the ball over winter, protected against their enemies by the thready mass.

How could such a relationship have arisen; what came first – gall wasp or wild rose able to produce threads in just that useful way?

But when you think about it, that is just another version of the old puzzle – what came first – bees or flowers? One cannot exist, today, without the other.

Opposite, *a* **Robin Pin Cushion gall**, *formed on the flower shoot of a wild rose bush.* Below top, *a* **housefly** *(3 × life size) and at* bottom *a* **bluebottle**. *They are all good examples of the close fit between animals and environment.*

Although animal species tend to be restricted in their life styles (they rarely live in unexpected places) change must be possible. Animals do 'colonise' new living places or 'habitats'. One example of change is that many of them live, as pests, with us in our own lives. Town pigeons are one example, but:

Look for houseflies, one is shown *above*. They are now typical of houses and homes, rather than of other urban habitats such as gardens.

Look for bluebottles, *right*. They are often seen in the garden, and may come into the house in autumn. However, if you do see them in the house, you will rarely see them land on food – they are looking for winter quarters to hibernate until spring. Would you consider them to be a pest?

Look for many other examples of animals which live alongside man, taking advantage of his activities. They range from starlings and house sparrows to clothes moths and weevils.

Look also for other animals which live in close association with man – yet are not counted as 'pests'.

Look for house martins; hedgehogs; and indeed garden birds.

In other words a 'pest' is a human definition, not a natural one!

Pests in the countryside

We also have a not very logical definition of pests in the countryside. Crows and rooks are scavengers, and play a useful role in nature. Yet because they will also take pheasants' eggs (pheasants are bred for shooting) the crows and rooks are often counted as pests.

Instinct

Most of the animal behaviour we see is called 'instinct' or 'innate behaviour' (it could be called 'habit'). The animal's instincts are as much a part of it, and as unchanging, as the shape of its body. They create its way of life.

The dunnock or hedge sparrow is feeding her nestlings *above*. The youngsters instinctively reach up and gape, and she instinctively feeds those gapes. She will feed only a reddish gape; a thrush will feed only a yellow one (that may cause problems for cuckoos).

Birds have an instinct to tidy up their nests, throwing away litter. One scientist ringed some nestlings and watched the parent bird instinctively pick up the metal rings and throw them out – with her brood attached.

Drives cause animals to behave in a certain way, and so instinct must be a number of drives, intricately woven together. Sometimes the end result of instinctive behaviour can seem (to us) to make 'good sense' – though sometimes we just do not understand what advantages it has. And life is, of course, very complex. Suppose that a predator approached the nest *above*. What 'instincts' would the bird follow? To fight back against the predator? To continue to feed her nestlings? To abandon her nestlings and fly to safety herself? How is the choice made?

The **hedge sparrow** opposite *responds to a red gape to feed its young.* Below, *another example of a 'trigger' – the* **red spot** *at the end of the* **gull's beak** *stimulates its nestlings to gape and beg for food. If it is missing, they do not.*

Unless a decision is made which overrules it, instinct is obeyed – it must be.

As a result, you can see many examples of instinctive behaviour which instead of protecting the animal, put it in great danger. The instincts gave the animal an advantage in certain circumstances – and the circumstances have changed.

And a good place to look is along roads.

Instincts have evolved during long periods of time. Cars are very recent things in the history of the world. Animals do not 'react' to them – in a car you can ride close to lions in a safari park, or close to pheasants and other birds along country lanes. The car is not part of their pattern of behaviour; they cannot 'recognize' it.

Look for hedgehogs, crushed by cars. When startled, their instinct is to stop still and begin to curl up as a defence. So the car strikes them.

Look at birds swooping across a road. Blackbirds will fly quickly out of a hedge, right in front of a car.

A rapid flight across past a predator would confuse it; but a car travelling at 50 mph cannot avoid hitting the bird.

Look for puzzling things which don't seem to have a reason, but which may be signals to trigger off a course of instinctive behaviour. A famous example is shown *right.* When the gull nestlings hatch, they will react to the red spot on the gull's beak, and gape when they see it – and *only* when they see it.

If by chance the parent's bill lacks that red spot, the nestlings do not gape, they do not receive food, and so they die from starvation.

Migration

Many species of animals migrate, butterflies as well as birds, and some of the distances they travel are immense. Butterflies cross wide stretches of open sea. Our swallows and house martins fly south in October, past the Sahara and deep into Africa. Arctic terns breed in northern Britain and Scandinavia, but fly to winter in the Antarctic – a round journey of more than 20,000 miles.

Migration is instinctive, and the puzzle lies in *how* the animals navigate. How do they find their way? Swallows and martins often seem to return to the same nest that they left.

Many theories have been proposed – that birds can use the stars as markers (experiments have been carried out in planetaria, which suggest that they can); or that they can also use the 'maps' made by polarised light (see page 5). And it is known that homing pigeons become confused when they cannot get a single glimpse of clear sky – when it is foggy, or totally overcast at night.

Some simple animals which live suspended in the sea have been found to have a 'compass' – an iron particle which directs their movements. Do the brains of homing birds contain iron particles to direct them?

The **swallow** opposite *is a familiar example of a migrating bird – often the same individuals seem to return to the very same nest after a round journey of thousands of miles.* Below, *migrants that winter here in Britain* – **Bewick swans**.

The investigation of how animals navigate is a job for scientists, but anybody can watch for migrants.

Look out for sudden 'plagues' of insects reaching Britain. In some years swarms of hoverflies or ladybirds arrive from Europe. It is a form of migration.

Look out for swallows and house martins. Without ringing, you would not be able to tell if they were the birds of the year before, returning. But what does their behaviour suggest. Are old nests always occupied? When are new nests built?

(Watch *how* new nests are built.)

Many birds migrate from further north to spend the winter in Britain; making up vast flocks on estuaries and mud flats. Indeed, almost the entire world populations of some wild geese winter in this country.

Look for Bewick swans, pictured *below*. They are smaller than other swans, and utter a yelping call when they fly. They come here from Siberia and are usually found concentrated in three places – the Derwent valley of North Yorkshire, the Ouse Washes in East Anglia and at Slimbridge in Gloucestershire. But they can be seen in other places as well.

Learning

Blue tits (*above*) visit gardens and often become tame. What can that mean?

It could signify two things at least. The bird has *lost* its 'fear' of us – it no longer sees our body and activities as a threat, or as a result of being tame it *gains* something – food!

We have already seen (on page 24) how instincts or innate behaviour can cause conflicts when danger threatens. The threatened bird had to make a choice – to continue to feed its nestlings, to fight back, or to flee. So does becoming tame mean that it makes a similar kind of choice?

But something is different. 'Tame' behaviour is *repeated* as a matter of habit. The bird comes to be fed again the next day, so that somehow it has 'remembered' the choice it made – even though instinct (as we have seen) is nothing to do with 'memory' of that kind. Instinct is *inflexible*.

But the blue tit has *learnt* to be tame. And now come the really fascinating puzzles: did the bird teach itself, or did it learn to be tame from another blue tit? Will it pass that tameness, or learning, down to its offspring? And if so, how?

And so we have come right back to a question that puzzled us earlier: we saw that birds seem to recognise that wasps are dangerous. How did they learn? Does every bird have to try a wasp and find out? Or do they inherit the learning of their parents?

The **blue tit** opposite *(and often the* **great tit** below*) are ideal birds to observe to discover clues about if, and if so how, birds can learn. You can set tasks, for example: if they peck milk-bottle tops, what happens if you paint the tops red?*

Blue tits seem to have learnt two or three different things – to eat coconuts, and to feed from a hanging bag of nuts. They also (as many people know!) peck the tops of milk bottles.

Why do they do it? Can we try to investigate milk-top pecking?

Look, for a start, at the way they open the top. Do they peck a hole, or tear it open? Do all the local tits use the same method? If so – then it could suggest that they learn from each other. Watch, and see what they do with the pieces – that may also be a clue.

Tits first started to attack milk-bottle tops in Southampton in 1929 – but soon birds all over the country did it – the habit (is that the word?) spread out from several centres until it was seen everywhere. Some even attack the bottles on the milk van.

So, it may be *learned* behaviour; but why do it at all? Of course, the bird gains the cream, but if it is learned from one to another, and so quickly and easily, then it must be rather similar to part of the bird's 'normal' behaviour.

Blue tits live on a diet of insects and other small animals. In a wood, their natural environment, they hunt their food in the leaves of the canopy far above ground (and in fact when they fly across a road, they often still fly exposed, high off the ground). But in the garden you will see them searching the rose bushes for greenfly.

However, a milk bottle does not seem to have much to do with a rose bush!

One theory is that inquisitiveness is an important part of learning. The animal explores, and if the result is of benefit, the animal repeats its original exploration. Tits need to be inquisitive to gain the supply of small animals they need for food; and maybe they will try anything that attracts their attention – even things which seem (to us) to be unlikely.

If you have never put up a coconut, try one in the garden. How quickly do the tits come to it? Try to observe if one learns from another to exploit this new food source.

Great tits (*below*) are also common garden visitors. Do they also peck at milk bottles, take nuts from the nut dispenser and eat coconut?

Groups

Many animals move, feed, or even 'rest' together in large numbers. Birds are often seen in flocks consisting of several different species. They may all be migrating at the same time and moving in the same direction at more or less the same speed. They may be birds of different species roosting together at night or vast nesting colonies of one single species. Small flocks of rooks (*above*) nest and feed together, making what is a small 'tribe'.

Do they communicate with each other, the birds in those flocks? Communication is interesting, and it may be one of the ways that instinct is over-ridden. If one bird of a flock sees a predator, it may give a warning call, which over-rides the 'safe' feelings of the others. They take avoiding action even though they themselves have not seen the predator.

The **rook** opposite *is a well-known social bird – its rookeries can be quite large.* Below, *another kind of animal grouping;* **massed waders**, *of many different species, attracted to the same spot by abundant supplies of shellfish.*

Look at flocks of starlings. Sometimes the roosting flocks can be immense. But what size are the *feeding* flocks? Are they always about the same size?

Look at the flocks flying, and wheeling and turning. How do they turn (nobody has yet been able to solve the problem of whether they follow a leader or not). You can, however, ask another interesting question – they often seem to wheel and turn for no good reason. Surely there *should* be a 'good reason'? If so, what can it be? Are they advertising the presence of the flock to smaller, scavenging groups who have not yet joined it?

Look at flocks of town pigeons. What happens when one is alarmed? What is their alarm signal?

Look at flocks of rooks. A rook is pictured *opposite*; it is often difficult to tell rooks from crows – because crows sometimes move in groups. However, rooks appear to have a longer bill (because it is bare of feathers at the base); also their thighs are covered with feathers, so that they seem to be wearing shaggy trousers.

You can try to analyse the wide range of noises which come from a rookery. The 'caws' presumably have a use in communication. Can you detect how?

More puzzles

Let's end this book where we started it – with stags. But a little later on in time – maybe a month or so from the placid scene on page 1. By now the stags have scraped the 'velvet' from their horns, annointed them with mud from a wallow, and are playing the mating game. They are fighting.

It is not a game, of course. But it is hardly in earnest either: little damage will be done to either stag. The loser will eventually realise the superior strength of the other, and retire. That mating behaviour makes a fascinating study.

After years of enquiry we are only just beginning to solve some of the mysteries of animal behaviour – we may be near to understanding how animals on migration navigate, for example. And, of course, part of the fascination will always lie with the fact that we too are animals.

So do our discoveries ever help us to understand the way *we* behave? Is our aggression the same, in origin, as the 'aggression' of stags at mating time? In a couple of months' time those stags above will be back sitting peaceably together as they were on page 1. What about our aggression?

Do animals have wars, or is their behaviour always a reaction, never an 'aggression' as we know it in the human world? Or is our 'aggression' nothing more than a reaction to triggers we ourselves cannot fathom?

Are we intelligent? Are animals intelligent? . . .

Yes, the investigation of behaviour is one of the most fascinating of all.